Planet
Earth

KINGFISHER

Kingfisher Publications Plc
New Penderel House
283–288 High Holborn
London WC1V 7HZ
www.kingfisherpub.com

First published by Kingfisher Publications Plc 2006
2 4 6 8 10 9 7 5 3 1
1TR/1205/PROSP/RNB/140MA/F

A CIP catalogue record for this book is available from the British Library.

ISBN 13: 978 0 7534 1242 8
ISBN 10: 0 7534 1242 X

Senior editor: Belinda Weber
Designer: Jack Clucas
Cover designer: Poppy Jenkins
Picture manager: Cee Weston-Baker
DTP co-ordinator: Lisa Hancock
Artwork archivist: Wendy Allison
Production controller: Jessamy Oldfield
Proofreader and indexer: Sheila Clewley

Printed in China

Acknowledgements
The publishers would like to thank the following for permission to reproduce their material. Every care has been taken
to trace copyright holders. However, if there have been unintentional omissions or failure to trace copyright holders,
we apologise and will, if informed, endeavour to make corrections in any future edition.
b = bottom, *c* = centre, *l* = left, *t* = top, *r* = right

Photographs: *cover* 1 Photolibrary.com; 2-3 Photolibrary.com; 4-5 Corbis Clay Perry; 6 Getty Imagebank; 7 Getty Stone; 8*bl*
Photolibrary.com; 8-9 Science Photo Library Roger Harris; 9*b* Photolibrary.com; 12*l* Corbis Rupak de Chowdhuri; 13*t* Photolibrary.com; 13*br*
Getty Photodisc; 12-13 Science Photo Library Pekka Parviainen; 16-17 Getty Imagebank; 16*cr* Photolibrary.com; 18*bl* Corbis NASA; 18-19
Getty Stone; 20-21 Corbis Tom Bean; 21*br* Getty AFP Yoshikazu Tsuno; 22-21 Corbis R.T. Holcomb; 23*tr* Corbis Charles & Josette Lenars; 24-
25 Photolibrary.com; 24*c* Corbis Galen Rowell; 25*tl* Getty Stone; 26*l* Corbis Michael Freeman; 26-27 Getty Imagebank; 27*t*
Photolibrary.com; 29 Corbis Audrey Gibson; 30-31 Getty Imagebank; 30*b* Corbis Robert Weight; 31*tr* Arcticphoto; 32-33 Corbis Yann
Arthus-Bertrand; 32*bl* Frank Lane Picture Agency Minden Pictures; 33*tl* Photolibrary.com; 34-35 Photolibrary.com; 34*tr* Frank Lane Picture
Agency Minden Pictures; 34*b* Getty Taxi; 35*bl* Corbis Craig Tuttle; 36-37 Frank Lane Picture Agency Minden Pictures; 36*b* Getty Stone; 37*tr*
Corbis Michael Yamashita; 38-39 Getty Digital Vision; 38*b* Getty Photodisc; 39*br* Corbis Tim Wright; 40-41 Getty Digital Vision; 40*bl* Getty
Photodisc; 41*c* Getty Photodisc; 48 Alamy Bryan & Cherry Alexander

Commissioned photography on pages 42–47 by Andy Crawford
Project-maker and photoshoot co-ordinator: Jo Connor
Thank you to models Alex Bandy, Alastair Carter, Tyler Gunning and Lauren Signist

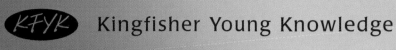

KFYK Kingfisher Young Knowledge

Planet
Earth

Deborah Chancellor

Contents

What is the Earth?

The Earth is a planet in space. It is one of nine planets that circle around the Sun in our Solar System. Seen from space, the Earth looks blue. This is because most of it is covered with oceans and seas.

Central America

The continents

The large areas of land are called continents. We can see the shape of the continents in photos taken from space. This continent is South America.

South America

The atmosphere

There is a blanket of gases around the Earth, called the atmosphere. White clouds swirl around in our planet's atmosphere.

Solar System – *the planets that revolve around the Sun*

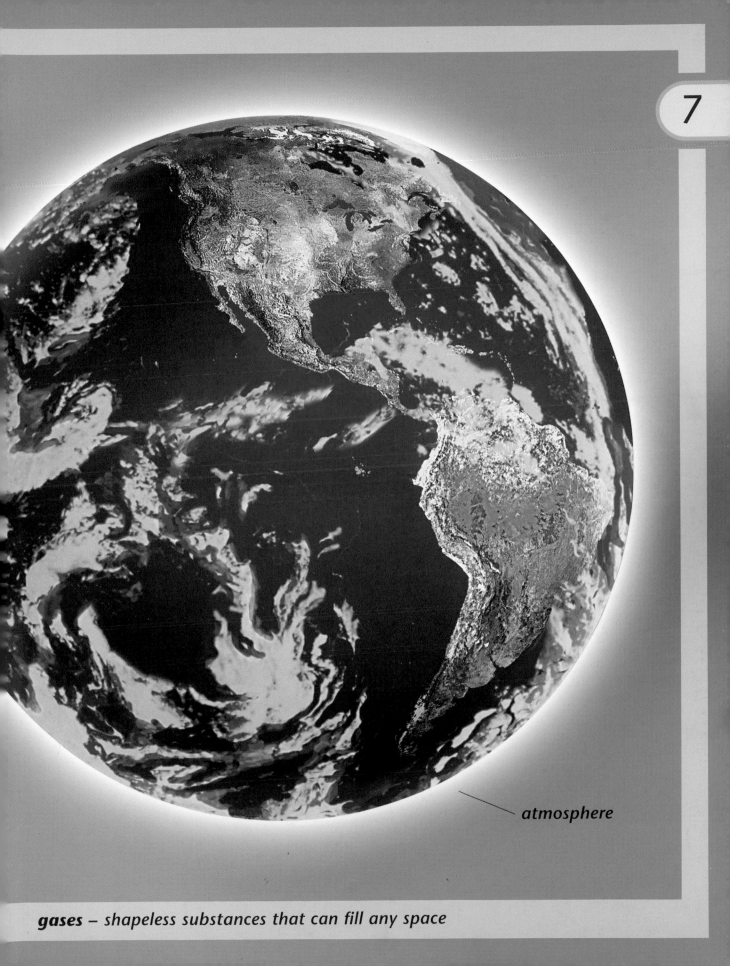

atmosphere

gases – *shapeless substances that can fill any space*

Inside the Earth

The Earth is a rocky planet. It is divided into three main parts, the crust, the mantle and the core. We live on the crust, which is a thin layer of solid rock. Not far under our feet, the rock is so hot that it is liquid.

crust

The crust

In some places under the sea, the Earth's crust is only six kilometres thick. Under most of the land, the crust is about 35 kilometres thick.

solid rock – *cool, hard rock*

The core

The core is the hottest part of the Earth. At its core, temperatures can reach up to 5,000°C.

mantle

inner core

outer core

The mantle

The hot rock in the Earth's mantle melts to become liquid. We can see molten rock when a volcano erupts.

molten rock – hot, liquid rock

The water cycle

The world's water is never used up. The Sun warms up sea water, turning it to water vapour. This vapour rises into the air, then comes down as rain. The rain then flows back to the sea. This is called the water cycle.

Sun heats sea water, making water vapour

water falls as rain

Water world

Most of the world's water is in the oceans. Only one per cent of all the world's water moves around in the water cycle.

vapour – water in the air

water falls
as rain

water vapour
rises to form
clouds

water collects
in rivers and
flows to the sea

Rain

Water vapour in clouds falls to the ground as rain. Some places get lots of rain. Mawsynram, in northern India, gets over 11 metres of rain every year. It is the wettest place in the world.

12 Weather and climate

Weather happens when the air around us changes. Air can be moving or still, hot or cold, wet or dry, or a mixture of these things. Water has a big part to play in the weather. Without it, there would be no clouds, rain or fog.

Tropical climate
The weather a place usually gets over a long time is called the climate. Climates vary in different parts of the world. In tropical places, the climate is hot and steamy.

pollution – harmful waste

Desert climate

In deserts, the climate is dry. On average, deserts have less than 2.5 centimetres of rain in a year. If all the rain comes at once, there are floods.

Trapping heat

Pollution in the air may trap some of the Sun's heat and stop it escaping back into space. As a result, our climate may be getting warmer every year.

tropical – an area around the middle of the Earth with very hot, dry weather

Clouds, rain and snow

Clouds are made of millions of tiny water droplets or ice crystals. Water droplets in clouds join together to make raindrops, and ice crystals combine to form snowflakes. Clouds come in many shapes and sizes.

Snow

Snowflakes usually melt on their way down to Earth. But if the air near the ground is freezing, we get snow.

droplets – very small drops of liquid

Different clouds

Low stratus clouds can bring rain. Fluffy cumulus clouds are seen on sunny days. High, wispy cirrus clouds are made of ice.

Cirrus cloud

Cumulus cloud

Thunder clouds

Cumulonimbus clouds are the biggest clouds of all. Some are taller than Earth's highest mountain, Mount Everest! They bring heavy rain, thunder and lightning.

Stratus cloud

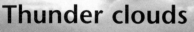
ice crystals – tiny pieces of ice

Wind

Wind is air that is moving around. It can be as gentle as a breeze, or as rough as a gale. Wind is made when the Sun warms the air to make it rise upwards. Cold air rushes in to fill the gap, making a wind blow.

When the wind blows

Wind travels at different speeds. A light breeze makes clouds drift across the sky. Stronger winds make trees sway, while very strong winds, called hurricanes, can cause lots of damage.

Air currents

Birds can glide along on rising currents of warm air. Seagulls hardly need to flap their wings to stay up high in the sky.

currents – movements in the same direction

Hurricanes and tornadoes

Hurricanes and tornadoes are dangerous wind storms. Hurricanes form over the sea, and when they reach land they can cause terrible damage. Tornadoes are powerful whirlwinds that form over land.

Hurricane

This satellite photo shows a hurricane in the Caribbean sea. It is heading for the coast of Florida, in the USA.

satellite photo – a photograph taken from a satellite orbiting the Earth

Twister

Tornadoes are also called twisters. Wind speed at the centre of a twister reaches up to 400 kilometres an hour – this is the fastest wind on Earth.

whirlwind – a strong wind that blows in a spiral

Earthquakes

The Earth's crust is made of many pieces, called plates. They are always sliding past or pushing up against each other. Sometimes this movement makes the ground split open, causing an earthquake.

How earthquakes happen

When two plates move suddenly, the ground judders and shakes, and deep cracks appear in the Earth's surface.

movement of plate

fault – *a break in the Earth's crust*

Fault line

This huge crack in the ground is the San Andreas Fault, in California, USA. Two of the Earth's plates grind past each other here. They have caused some massive earthquakes.

Earthquake drill

Earthquakes are quite common in some places. In Japan, school children wear protective hats when practising what to do if there is an earthquake.

drill – a repeated practice or exercise

Volcanoes

A volcano is a mountain made out of molten rock, called lava. This molten rock comes from deep under the ground, and it forces its way up through cracks and weak points in the Earth's crust. Volcanoes can form on land or deep under the ocean.

Inside a volcano

Molten rock, called magma, collects in a chamber. When the volcano erupts, it is forced upwards, through a vent.

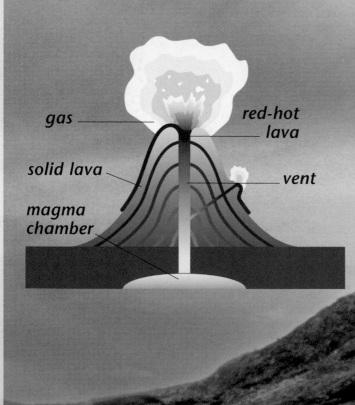

gas

red-hot lava

solid lava

vent

magma chamber

lava – molten rock on the Earth's surface

Bubbling mud

The land around volcanoes becomes very hot. Pools of mud or water bubble and boil on the Earth's surface.

Giant volcano

The biggest active volcano in the world is in Hawaii. This lava flow is from the Mauna Ulu crater there.

magma – molten rock underground

Mountains

Mountains form over millions of years. They are made when two plates under the Earth's crust push together, forcing up huge folds of rock. As the mountain is pushed upwards, ice, wind and weather wear it down. This is called erosion.

plates – large areas of land that 'float' on the molten rock underneath

The Alps

The Alps in Europe are a few million years old. Young mountains have jagged peaks. The weather has not had time to smooth down the sharp edges of rock.

The Himalayas

The 14 tallest mountains in the world are in the Himalayas, in Asia. These mountains are over 50 million years old.

peak – the top of a mountain

Rivers and lakes

All rivers carry water to the sea. Some are so powerful they change the shape of the land they pass through. They carry rocks and mud along with them, cutting deep valleys and gorges as they go.

Lakes

Lakes are large pools of water surrounded by land. They can form in volcanic craters or in valleys made by movements of the Earth's crust.

A river's journey

A river begins its journey in high ground, where it flows quickly downhill. When a river enters a valley, it flows slowly in bends, called meanders.

valleys – *areas of low land*

Grand Canyon

The Colorado river in the USA has carved out the deepest gorge in the world. The fast-moving waters have worn away the rock to help create the amazing Grand Canyon.

gorges – *valleys with steep sides*

The oceans

Oceans cover most of planet Earth. They are deeper in some parts than in others. This is because the ocean floor is not flat. There are mountains, valleys, plains and deep trenches under the sea.

Big blue sea

The five main oceans are the Arctic, Atlantic, Pacific, Indian and Southern oceans. The largest of these oceans is the Pacific.

volcano

shipwreck

deep trench

trenches – *long, narrow valleys*

Low tide

At low tide, rockpools can be found on rocky beaches. The pools are covered over again at high tide.

Islands

Some underwater mountains and volcanoes are so tall, they rise above the surface of the water. Many islands are actually the tips of underwater mountains.

mountain range

island

The poles

The North Pole is in the middle of the Arctic ocean. This is a frozen ocean, surrounded by the world's most northern lands. The South Pole is at the heart of Antarctica. Most of this continent is covered with thick ice.

Antarctic science

Antarctica is the coldest and windiest continent. The only people who live there are scientists, who work in research stations.

glaciers – moving rivers of ice

Icebergs

In Antarctica and the Arctic, icebergs break away from ice sheets or glaciers and float in the icy ocean. We see only a tiny part of an iceberg – the rest is hidden under water.

Northern Lights

The Northern Lights, or aurora borealis, can be seen in northern Canada, Alaska and Scandinavia. The spectacular display takes place high up in the atmosphere.

atmosphere – air around the Earth

Deserts

Deserts are the driest places on Earth, because it hardly ever rains. Some deserts are sandy, and others are rocky. Some are very hot, while others are freezing cold in winter.

Desert plants

Cactus plants grow in American deserts. They can live for a long time without rain, because they store water in their thick stems. Some birds make their homes in cactus stems.

cactus – a plant that can grow in places with little rain

Wind erosion

Deserts can be windy places. Wind blasts sand at tall rocks, gradually wearing them away. The rocks in Monument Valley, USA, show how wind can change the landscape in a desert.

Biggest desert

The Sahara, in northern Africa, is the biggest desert in the world. It contains the world's tallest sand dunes, which are up to 430 metres high and five kilometres long.

sand dunes – *big hills of sand that are formed by the wind*

Forests

A forest is a large area of land covered in trees. About a fifth of the world is covered with forest. In the past, forests grew over much more of the planet, but people have cut many of the trees down.

Deciduous

Trees that lose leaves in the winter are called deciduous. The leaves change colour and fall off the trees in the autumn.

Rainforest

Rainforests grow in hot countries where there is a lot of rain. The wettest rainforests have over ten metres of rainfall a year.

Evergreen

Big forests of evergreen trees grow in northern parts of the world. Evergreen trees do not lose their leaves in winter. Their branches slope down, so the snow slides off them.

rainforest – a thick, tropical woodland

Life on Earth

Earth may be the only planet in the universe that can support life. Our planet's oxygen, and water in the oceans, are vital for living things to survive.

Rainforest life

There are many millions of types, or species, of animals and plants on Earth. Tropical rainforests are home to more than half of the world's plant and animal species.

When life began

Scientists believe that life on Earth began over 3.5 billion years ago. It has been slowly changing, or evolving, ever since. Remains of ancient creatures tell us a lot about life a very long time ago.

In the sea

The oceans were home to the world's first animals. Some ocean species, such as sea turtles, are over 200 million years old.

oxygen – *one of the gases in air*

Earth's riches

Many of the Earth's natural riches are hidden deep under the ground. Fossil fuels are found in rocks, thousands of metres below the Earth's surface. They are made from the remains of ancient plants and animals.

Minerals

Rocks are made from minerals. Rare minerals, such as the diamonds and rubies in this crown, are called gems.

mineral – a hard, natural substance

Oil and gas

Oil and gas are fossil fuels, which are pumped up from holes drilled into the Earth's crust. They are found in places that are, or once were, under the sea.

Coal

Coal is a fossil fuel that is burned in huge amounts to make electricity. It is dug out from deep underground mines.

crust – *a layer of rock around the Earth*

Looking after the Earth

The Earth gives us food, water and air to breathe. Sadly, people have not looked after it, and many places are now polluted. Lots of plants and animals have died out, or soon will do. We must all help make the Earth a cleaner place.

Saving forests

Trees help keep the air clean, and provide shelter for many different animals. People must stop cutting down so many forests, and plant more trees.

polluted – made dirty

Recycling

We can make new things from old materials. This is called recycling. Bottles, cans, paper, plastic and aluminium foil can all be recycled.

New energy

Scientists are developing new forms of energy that do not pollute. Many of the forces of nature, such as the wind, can be used to make electricity.

Make a volcano

Understanding eruptions

There are about 700 active volcanoes in the world today. When a volcano erupts, huge underground pressures force liquid rock up into the air. You can make your own volcano with some simple materials. In your volcano, sodium bicarbonate mixes with vinegar to make carbon dioxide gas.

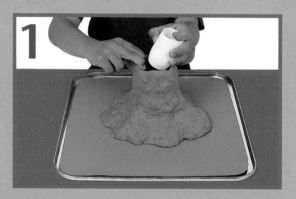

Using the clay, make a hollow volcano and place it on the tray. Slide the plastic bottle inside.

You will need
- Modelling clay
- Baking tray
- Small, plastic bottle with the top sliced off
- Sodium bicarbonate
- Funnel
- Vinegar
- Red food colouring

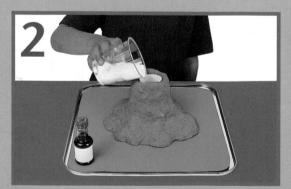

Half fill the bottle with sodium bicarbonate. You may need to use the funnel for this.

3

Place your volcano and baking tray on a flat surface. You can take it outside if you prefer.

4

Mix the vinegar with the food colouring. Pour them into the bottle using the funnel.

Stand back and watch your volcano erupt!

Make a rain gauge

Measuring rain

There is an easy way to measure how much rain falls during a shower. Put your rain gauge out in the open. When the shower is over, open the lid and collect the rain water in a measuring jug. Note down how much rain fell.

You will need

- Large plastic bottle
- Scissors
- Elastic bands
- Garden pole or stick
- Measuring jug
- Pen and paper

Cut a section off the plastic bottle, using the scissors. You may need to ask an adult to help you.

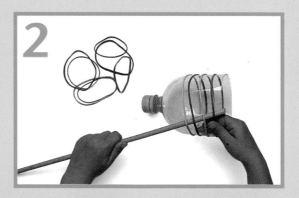

Put elastic bands around the bottle. Slide the pole under the bands and position the bottle with the screw cap facing down, to catch any raindrops.

Make a windmill

Spinning sails

You cannot see the wind, but you can watch what it does. Make a windmill and see how the wind blows it around.

You will need
- 2 squares of coloured card
- Pencil and ruler
- Scissors and sticky tape
- Drawing pins and wooden rod

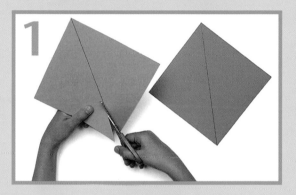

Draw a line across each paper square, from one corner to the other. Cut along this line to make two triangles.

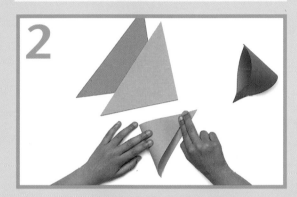

You will now have four triangles. Fold each of the triangles in half, fixing the corners together with sticky tape.

Place the corners of the four triangles on top of each other. Ask an adult to help you pin them on to the wooden rod.

Forest habitat

Make your own forest

The place in which an animal lives
is called a habitat. Everything an
animal needs to survive can be
found in its habitat, for example,
food and shelter. There are many
different kinds of animal habitat on
Earth. You can make a model of a
forest habitat with craft materials.

*tree template to
draw around*

You will need
- Big shoebox
- Poster paints and brush
- Pencil
- Tracing paper
- Card
- Scissors
- Glue
- Modelling clay
- Plant material – leaves,
  grass or twigs
- Toy forest animals

1 Paint inside your shoebox, using brown paint for the ground, green for the grass and blue for the sky.

2 Use the template on the opposite page to draw some trees on card. Cut out the plant shapes.

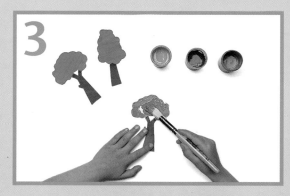

3 Paint the trees. When they are dry, fix them inside the box with glue or clay. Scatter the plant material on the ground and arrange your animals in their new home.

Index